Table of Contents

POLYMERASE CHAIN REACTIONS:

CLONING:

PART I
Polymerase Chain Reaction (PCR)

Gabriella de Souza

SPRING 2014

Polymerase Chain Reaction (PCR): A method to amplify a single or a few copies of a piece of DNA across several orders of magnitude, generating thousands to millions of copies of a particular DNA sequence.

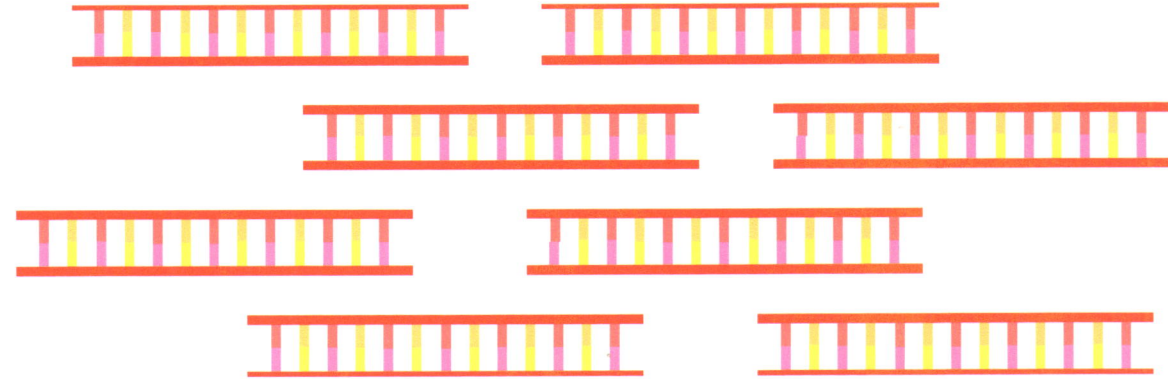

Think of PCR the same way you think of a copy machine:

Photo Copier Items	PCR Components
The Book	The entire Genome (called the DNA template)
A Page	The target DNA sequence
A Bookmark	The forward and reverse primers that "mark" the specific DNA fragment.
The Copy Machine	The enzyme that copies the DNA (called a Polymerase)
Paper and Toner	The four nucleotides that make up DNA. (In this case: the master mix)

A List of the materials you will need:
- 25μL of Master Mix
- Cold Block for Master Mix and PCR tubes
- 1μL of Forward Primer
- 1μL of Reverse Primer
- 1μL of DNA Template

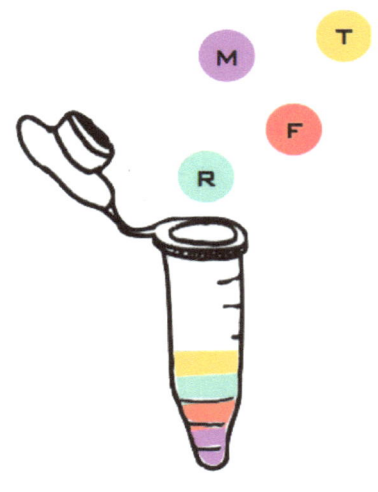

PREPARING THE PCR:

The preparation begins by thawing out the master mix. Next, place the appropriate number of PCR tubes in the cold block and cover them with the cold block lid. To each PCR tube, add 25μL of Master Mix, 1μL of the forward primer, 1μL of the reverse primer, 1μL of the DNA template, and 22μL of the nuclease free water. After, mix the contents of each tube gently with a pipette. Cap the tubes tightly; They are now ready for the Thermocycler.

PCR is broken down into three steps with the addition of an initial denaturation. These include: Denaturation, Annealing, and Extension. These steps are repeated from 28-35 times. Each cycle is about 5 minutes; therefore, a large quantity of DNA can be produced in a matter of hours.

Initial Denaturation:
This step prepares the DNA for step 1, and ensures that the DNA is completely denatured by the end of step 1. The DNA is heated to 95°C.s

STEP 1: Denaturation
The DNA is heated usually to 95°C so that the double helix is separated.

DNA SEQUENCE   TARGET DNA   NUCLEIC BASE PAIRS

INITIAL DENATURATION

95°C

STEP 1: DENATURATION   STEP 2:

ANNEALING   95°C

The forward and reverse primers attach to their designated ends of the target DNA. The temperature for this step varies—depending on the type of primer used.

STEP 3: EXTENSION
In this step, "DNA synthesis" occurs. The DNA polymerase attaches to the primers. During this step, the DNA is heated to approximately 72°C. The DNA polymerase then extends and "synthesizes" to create a new strand.

STEP 2: ANNEALING

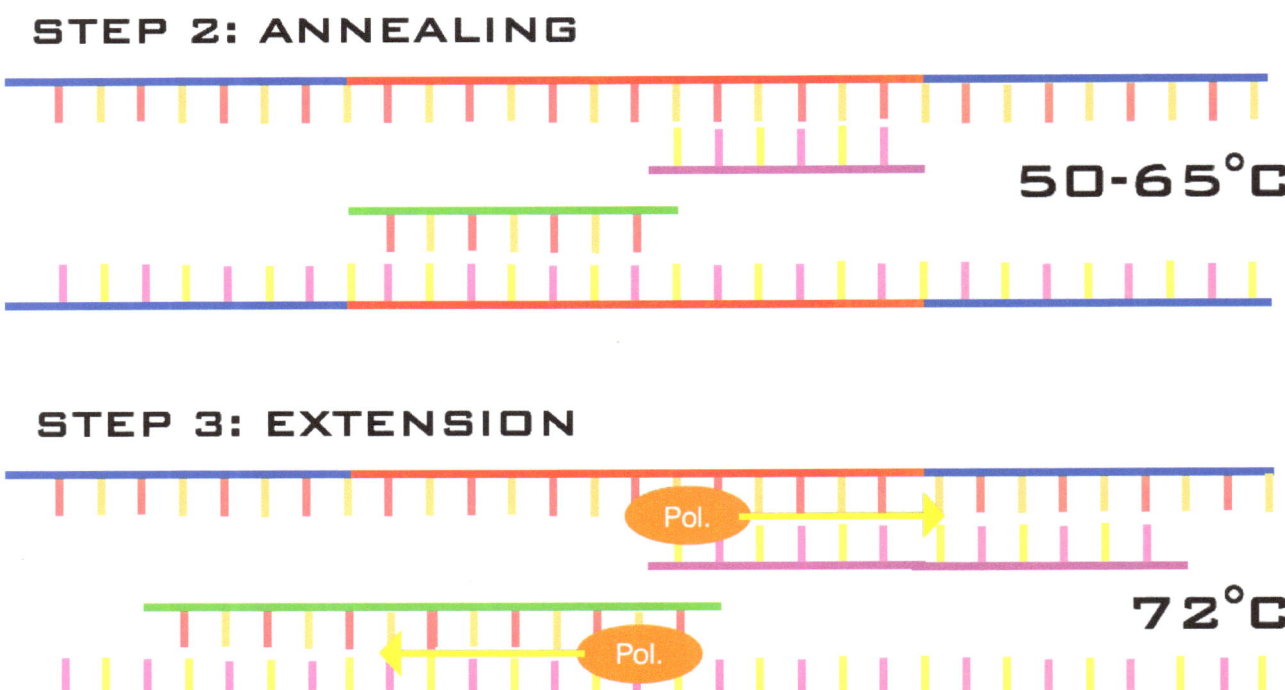

50-65°C

STEP 3: EXTENSION

72°C

As mentioned before, the cycles repeat up to 35 times. For each cycle, the strand of Target DNA will copy itself.

Think of it this way:
Starfish reproduce using "asexual reproduction". Each starfish makes another starfish.

Polymerase Chain Reaction Steps In Review:

- Initial Denaturation: "initially" peels apart the DNA.
- Denaturation: Continues to separate the double Helix.
- Annealing: Binds the forward and reverse primers.
- Extension: Synthesizes the DNA, creates a new strand.
- Repeat: These cycles repeat up to 35 times in the thermocycler.

STEP 1: DENATURATION

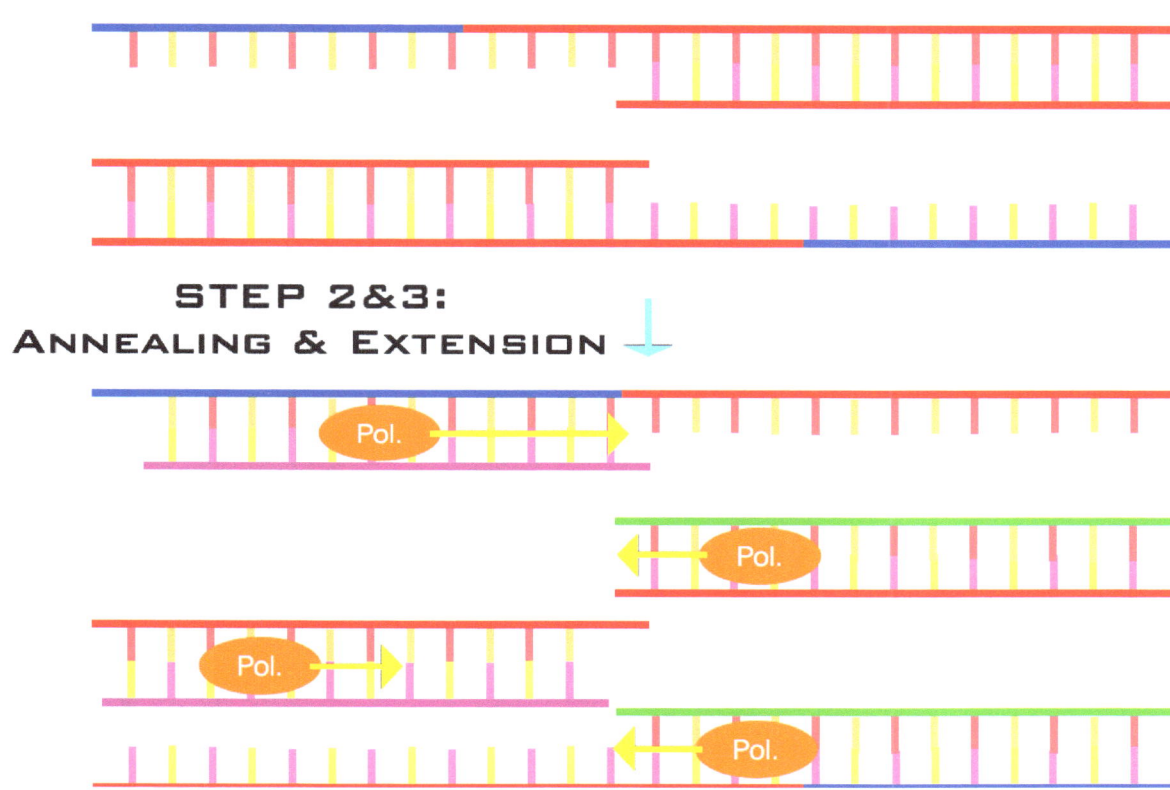

STEP 2&3:
ANNEALING & EXTENSION

Glossary:

Asexual Reproduction: An organism capable of asexual reproduction is able to produce offspring in the absence of a mate. In asexual reproduction, the offspring is a clone of the parent and therefore results in low genetic variation in the species as a whole.

Cycle: All the steps of PCR form a cycle.

DNA (Deoxyribose Nucleic Acid): A double-stranded nucleic acid that contains the genetic information for cell growth, division, and function.

DNA Template: A structure that in some direct physical process can cause the patterning of a second structure, usually complementary to it in some sense.

Nucleotides: The basic building block of nucleic acids, such as DNA and RNA. It is an organic compound made up of nitrogenous base, a sugar, and a phosphate group.

Pol. (Polymerase): The enzyme that catalyzes the reaction for DNA synthesis.

Primers: A primer is a strand of nucleic acid that serves as a starting point for DNA synthesis.

Step: Each individual action within PCR. For example, Denaturation, Annealing, or Extension.

Target DNA Sequence: The desired DNA sequence of interest.

Thermocycler: The machine in which each step and cycle occurs. This machine heats and cools the PCR tubes in order for reactions to occur.

PART II
Polymerase Chain Reaction Protocol

Gabriella de Souza
Dr. Ryan G. Rhodes

FALL 2013

Preparation of Polymerase Chain Reactions (PCR)

Background and Purpose:

The Polymerase Chain Reaction (PCR) is a molecular biology method that was developed by Kary Mullis in 1983 and is used to amplify a specific fragment of DNA. PCR reactions contain: 1) DNA polymerase, 2) deoxyribonucleoside triphosphates (dNTPs: dATP, dTTP, dCTP and dGTP), 3) a buffer containing $MgCl_2$, 4) a forward primer, 5) a reverse primer, and 6) a DNA template. The PCR reactions described here utilize 2x Master Mixes that are commercially available and simplify reaction setup. Specifically, each 2x Master Mix contains the DNA polymerase, dNTPs and buffer and all that needs to be added to the Master Mix are the primers, template and the appropriate volume of nuclease-free water. Several different Master Mixes are listed below and are used for a variety of different purposes. For example, when a DNA fragment will be used in cloning reactions it is important to use a DNA polymerase with high fidelity (low frequency of mutation). On the other hand, sometimes you may only be evaluating the product of a PCR reaction by gel electrophoresis (eg. colony PCR), in this case fidelity does not matter because you are only interested in the size of the product. Therefore, you can use any Master Mix available (typically the cheapest).

Note that designing primers is described in a separate protocol, but using primers with similar melting temperatures usually provides the best results. The template added to a PCR reaction may be purified DNA or a crude extract. For example, purified DNA (genomic DNA or plasmid DNA) works best when available or appropriate, but there are times when using a crude extract such as the supernatant from lysed cells is necessary (eg. colony PCR). Once the PCR reaction setup is complete, the reactions are placed in a thermal cycler which is programmed to cycle through the following stages: 1) Initial denaturation; 2) 30-35 cycles of denaturation, annealing, and extension; and 3) final extension. Times and temperatures for each stage may vary depending on the master mix used, the melting temperature of the primers, and the length of the DNA fragment.

Materials:

- 2x Master Mix (NOTE: keep on ice or cold block at ALL times)
    - Q5 High Fidelity 2x Master Mix (New England Biolabs M0492L; use for cloning) OR

- o Quick-Load Taq 2x Master Mix (New England Biolabs; use for colony PCR) OR
- o Premix Taq: ExTaq 2x Master Mix (TAKARA; use for cloning or colony PCR)
- Cold block or ice for 2x Master Mix
- Cold block for 0.2mL PCR tubes
- Forward Primer (10 μM)
- Reverse Primer (10 μM)
- Template DNA

Procedure:

1. Thaw the 2x appropriate 2x Master Mix and mix briefly with a 1000 μL micropipette
2. Place the appropriate number of 0.2 mL PCR tubes in the cold block and cover

Note: The following steps assume a 50 μL final volume

3. Add 25 μL of the 2x Master Mix to each PCR tube
4. Add 1.0 μL of the 10 μM forward primer (final concentration is 200 nM)
5. Add 1.0 μL of the 10 μM reverse primer (final concentration is 200 nM)
6. Add 1.0 μL of purified DNA template (if using a crude template add 2.0 μL)
7. Add 22 μL of nuclease-free water (water is added to bring the volume to 50 μL)
8. Mix gently with pipette tip
9. Centrifuge if necessary to pull contents to the bottom of the tube
10. Cap tubes tightly and transfer to the thermal cycler
11. Confirm the correct thermal cycle program and run (see below for thermal cycles using master mixes listed in this protocol)

Note: See DNA electrophoresis protocol for analysis of PCR reaction

Thermal Cycles for Q5 High-Fidelity 2XMaster Mix or Premix Taq: Ex Taq

Number of Cycles	Step	Temp.	Time
1 cycle	Initial uration	98°C	10 sec
35 cycles	Denaturation	98°C	10 sec
	Annealing	55°C*	30 sec
	Extension	72°C	1 min**
1 cycle	Final Extension	72°C	5 min
--	Hold	10°C	∞

Thermo Cycle (Quick-Load Taq 2X Master Mix)

Number of Cycles	Step	Temp.	Time
1 cycle	Initial uration	95°C	10 sec
35 cycles	Denaturation	95°C	10 sec
	Annealing	55°C*	30 sec
	Extension	68°C	1 min**
1 cycle	Final Extension	68°C	5 min
--	Hold	10°C	∞

*Annealing temperature will vary depending on melting temp of the primers (set 2 degrees below melting temp)

**Extension time will vary depending on length of product (use 1 min per kb of DNA being amplified)

PART III

Cloning: Legality, Religious Views, and Benefits

Gabriella A. de Souza

SPRING 2014

Outline

<u>Thesis:</u> Through funding and political support for non-embryonic stem cell research, scientists can produce pluripotent stem cells without the use of embryonic material, thus accommodating the legal and religious views against cloning, while maintaining its benefits.

I.   The use of cloning: legality, history, and moral implications

    A. Current cloning laws in The United States

        1. Laws by state

    B. The history of the misconception

        1. Influence of media

    C. Lack of cloning laws in the past

        1. Past experiments and studies

II.   Religious views on cloning: ethical issues

    A. Christianity

        1. Christian views against all types of cloning

    B. Theravada Buddhism

        1. Buddhist objections to reproductive cloning

    C. Islam

        1. Islamic objections to reproductive cloning

III.   Cloning: Procedures and their benefits

    A. Current organ supply

    B. New methods for pluripotent stem cell production

    C. Overall benefits and disadvantages of cloning

The church and state frown upon cloning due to its science fiction repute. Even society has turned its back on the potential of therapeutic and reproductive cloning. Commonly, when a person hears the term "cloning", their minds immediately wander to thoughts of replicating entire organisms, a process known as "reproductive cloning". Gregory E. Pence expands on this idea saying, "A long legacy in science fiction novels and movies make the word 'cloning' so fraught with bad connotations that it can hardly be used in any discussion that purports to be impartial" (Pence 28). This widespread misconception blinds society from knowing the definition of cloning. It has been proved that reproductive cloning is possible, but constructing identical fetuses is not the sole purpose of cloning. On a daily basis, cloning is used in thousands of microbiology labs during a process called Polymerase Chain Reactions or "PCR". This process allows researchers to magnify DNA [genetic material; Deoxyribonucleic Acid] by producing millions of copies of a desired DNA sequence in only a matter of hours. That DNA can then be analyzed and further manipulated. Another example of cloning can be found in bacterial colonization, when bacteria colonize and reproduce themselves. Persons lacking knowledge of the various types of cloning are under the misconception that cloning is strictly for reproducing identical organisms.

This paper will be discussing two important cloning methods; therapeutic cloning and reproductive cloning. Therapeutic cloning is performed for the purpose of medical treatment. This sometimes involves the use of one of the most controversial materials of the 21st century: embryonic stem cells. The reason these cells are so heavily debated is because they come from the embryo or zygote[1] of a living organism. This process defies many ethical views and policies placed by legislature and various religions. In comparison, reproductive cloning needs these stem cells to produce life. Studies have recently been conducted, claiming that the same pluripotent stem cells can be man-made rather than having to extract them from an organism's embryo. The lack of government funding for stem cell research is holding back medical and scientific advancements. Through funding and political support for non-embryonic stem cell research, scientists can produce pluripotent stem cells without the use of embryonic material, thus accommodating the legal and religious views against cloning, while maintaining its benefits.

Throughout history, the government has oppressed the potential of therapeutic cloning and reproductive cloning because of society's

---

[1] Zygote: the product of the union of male and female sex cells; before the creation of a fetus.
[2] Nuclei, Nucleus: the center or "brain" of a cell containing all the DNA and genetic information.

16

misconception and lack of experience in this realm of science. Due to the ethical ambiguity that comes from this form of science, laws have been established to prevent cloning research. Despite this, the laws between states in the U.S. vary, allowing research to be conducted. In 2011, the Bioethics Defense Fund gathered data based upon laws that could apply to cloning, and complied it into a fifty state survey. The survey says: eight states have prohibited cloning for any purpose. To expand, the law bans the use, creation, and research of human embryos and embryonic stem cells. There are fifteen states that allow the use of cloned embryos as long as they are destroyed in research. In translation, these states permit the study and creation of human embryos and or embryonic stem cells; although they demand that the embryo created must be destroyed in research (Nikas 2). The remaining states have not enforced any laws pertaining to cloning research; therefore, there are no specific laws that could prohibit cloning studies in these states. It is also important to identify the possible causes for creating these research limitations. One cause may have been the study of cloning by The President's Council on Bioethics. The council prepared an argument against cloning, stating that, "Current cloning technology has an extremely high failure rate" and that "there is no way to safely increase the success rate without sacrificing human embryos in the process" (Roleff 16). This is just one example that people are being misinformed of cloning and its benefits. The misconception has caused the government to prevent scientists from conducting valuable research.

More so, there has been evidence of the media corrupting opinions on cloning research, causing the rate of biotechnology advancements to decrease. Society has paired therapeutic cloning and reproductive cloning together causing therapeutic cloning to be drenched with disparagement; in other words: suggesting that they are the same process. However, it is known that therapeutic cloning and reproductive cloning are quite different. Dr. Candace Gauthier, a professor from the University of North Carolina at Wilmington, stated that, "[She feels] like the moral and political controversies are attached to reproductive cloning and it places a dark shadow over therapeutic cloning, giving people the idea that therapeutic cloning is a negative thing" (Gauthier). Furthermore, within media and social networking, movies and television shows are influencing the public and further corrupting their definitions of therapeutic and reproductive cloning. For example, the book and film, "Never let me go", which focuses on a group of children who have been cloned exclusively for the purpose of organ donation. The children are told that their

18

only purpose is for organ donation and that they will "complete" or die in their early adulthood. These cloned donors are not allowed to have relationships or produce children, as the schoolmasters believe they have no soul. This story illustrates the negative use of cloning, and persuades viewers that reproductive cloning is the only type of cloning. Although, it is known that cloning can also be used for the purpose of medical treatment; otherwise known as therapeutic cloning. Similarly, the television show, "Orphan Black", features a woman who assumes the identity of her clone after witnessing the latter's suicide. The series raises issues about moral and ethical implications of human cloning as well as its impact on issues of personal identity. This is evidence that media creates great amounts of controversy in the scientific research world, and therefore fuels the corruption of the public opinion.

In history, because of the misconception, the cloning process seemed science fictional; therefore, in other countries no laws were made to limit the scientists' research. In the year of 1938, German embryologist, Hans Spemann articulated the principles of modern cloning. Spemann contemplated the idea of animals being "replicated by transferring the genetic material [DNA] of differentiated (somatic) body cells—skin cells, for example—to egg cells whose nuclei[2] had been removed" (Hansen 11). In 1984, Steen Willadsen of Denmark used this method—which became known as nuclear cell transfer—to reproductively clone a sheep embryo. This process was passed down and later used in 1996 by Ian Wilmut to develop the famously cloned sheep, Dolly. Wilmut and his research group used a cell's nucleus from an adult Finn-Dorset sheep and injected it into an unfertilized oocyte[3] that had its nucleus removed. The nuclear transfer method that Spemann developed is now used in modern processes for therapeutic cloning and reproductive cloning. Furthermore, due to the lack of laws against cloning in other countries, "scientists who wanted to do this research have left [America]" says professor Dr. Candace Gauthier. The American legislature is causing what is left of scientific advancements to deteriorate. Overall, due to the newly established cloning laws, there has been a large decline in biotechnology advancements.

Not only is the government concerned with cloning, various religious groups including Christianity, Islam, and Theravada Buddhism, also have major concerns. Moreover, Christianity is one of the most widely practiced

---

[2] Nuclei, Nucleus: the center or "brain" of a cell containing all the DNA and genetic information.
[3] Oocyte: a female gametocyte or germ cell involved in reproduction (used to produce a zygote).

religions in the world; it also has been known to be one of the most influential religions in history. Traditional Christians do not support the act of cloning whatsoever, and they believe that it is against the religion entirely (Weasel & Jensen 5). When pastors preach to the congregation, they emphasize the word "DNA" to indicate that it is one's "God-given Identity" (Weasel & Jensen 8). This indication symbolizes the uniqueness and individuality that God has given all of his followers. Because of the evolution of Christianity and Christian beliefs, there are some Christians who feel that the idea of therapeutic cloning is beneficial to medicine, as long as it is for a medical purpose (Weasel & Jensen). Although, if embryonic stem cells are used, Christians would no longer support the method for therapeutic cloning, as harming human embryos is against Christian beliefs. The major conflict that occurs between cloning and the Christian doctrine is the idea that overstepping the natural order is too similar to "playing God" (Weasel & Jensen 11). Although the concept of cloning is against Christian axioms, most contemporary followers accept the idea of therapeutic cloning, but not reproductive cloning.

In contrast, Theravada Buddhists reject the idea of harming or killing living beings, but accept the ideas of reproductive and therapeutic cloning.

Harming living organisms is against natural order, as Buddhists claim (Gomes 49). Though cloning can sometimes involve embryonic stem cells, if society begins to support methods of pluripotent cell research that lack the use of an embryo, this realm of science can advance towards the reality of safely cloning cells for medical purposes. In order to respect Buddhist axioms, the cloning method used must agree with the five precepts known as Pañca Sīla. Theravada Buddhists must live their lives in accordance to the Pañca Sīla. The first precept is "to refrain from killing living beings". The second is "to refrain from taking what is not given"; meaning to not steal, or be an accessory to theft. The third precept is "to refrain from sexual misconduct"; this includes any practice of sexual activity. The fourth is "to refrain from false speech". Finally, the fifth precept is "to refrain from drugs and drinks that tend to cloud the mind"; meaning to not partake in activity involving drugs or alcohol (Gomes 47). It is very clear that only one of these precepts involves cloning or has a possibility of causing conflict with the act of cloning. This precept would be the first, which was "to refrain from killing living beings". If scientists can find a method of cloning that does not involve the harm or sacrifice of living beings, they can satisfy the concerns of the Buddhist people. In conclusion, Buddhists must refrain from harming life, and if there is

22

support for cloning and its research, techniques can be developed that do not involve the harming of living organisms.

Similar to Christianity, Islam also has concerns with reproductive cloning, as it suggests that man is "playing God". Even though Islamic people do not accept the replication of man through reproductive cloning, they praise the ideas that include therapeutic cloning; for example, cloning organs, muscles, etc. In Islam, an embryo, even when it is first developing, has the right to live; however, there are different stages of "ensoulment[4]" throughout the production. The Holy Quran mentions the stages of life in chapter 23, verses 12-14: "We created man of an extraction of clay, then We set him a drop in a safe lodging, then We created of the drop a clot, then We created of the clot a tissue, then We created of the tissue bones, then We covered the bones in flesh; therefore We produced it an another creature. So blessed be God, the Best of Creators" (Larijani & Zahedi). Therefore, because of the identification of the stages of life, some manipulation to a human embryo is allowed—as long as it does not bypass the Muslim law. Knowing this, therapeutic cloning, even if it involves the use of an embryo, should be

---

[4] Ensoulment: is a philosophical or religious concept referring to the moment at which a human being gains a soul.

accepted by the laws of Islam. The standing of the embryo has been declared based on the opinions of the Muslim scholars of Sunni and Shi'a; "it [ensoulment] takes place about the end of the fourth month (120 days after fertilization)" (Larijani & Zahedi). Thus, in the religion of Islam, reproductive cloning is not allowed, while therapeutic cloning methods will be considered.

Still, therapeutic cloning and reproductive cloning have great potential, and can benefit society if properly funded and supported. For example, therapeutic cloning—with the use of pluripotent stem cells—could help repair damaged cells which would increase organ health and quality. Currently, there are an extremely low amount of transplant organs. It is said that, "one patient is added to the transplant waiting list every 15 minutes" (Abouna 34). The majority of the patients who are added to these lists die during their wait for an applicable donor. Not only is scarcity an issue, but organ rejection can also occur. The immune system targets the transplanted organ as a foreign object and attempts to remove it. In order for a patient to have a successful transplant, they are prescribed immunosuppressant medications that "suppress" the immune system so that the organ can adjust to its new environment. However, there are some cases in which the patients, even though they are taking the prescribed medications, continue to have organ rejection; thus leading to organ failure. Sally Liu of Duke University believes that there is one solution to this immune rejection problem: somatic cell nuclear transfer; otherwise known as therapeutic cloning. This method can be used to create organs that would be less likely to fail in the new patient's body. These tissues would be derived from the patient and therefore be fully accepted by the immune system—allowing the patient to fully recover. If the government implements funding for therapeutic cloning research, there will not be such a strong need for organ donors as the ability to create these organs or repair damaged tissue will become a reality. Repairing the damaged cells in the organ won't only increase the health of the surrounding tissue, but it also has the ability to treat diseases such as Alzheimer's, Parkinson's, heart diseases, or diabetes; as shown in Table 1.

In addition, the scarcity and rejection of transplant organs won't be as concerning. As a result, transplant organs are scarce and patients in need of a transplant are likely to experience organ failure or organ adjustment difficulties after their procedure. Moreover, with the use of therapeutic cloning these issues would no longer be a concern, as the patients would be receiving an organ that is genetically identical to their own.

There are various procedures in which scientists can extract the

pluripotent cells that are crucial to therapeutic and reproductive cloning.

Unfortunately, the majority of these procedures are known to involve the harm

or destruction of human embryos as previously mentioned. However, as of

January 2014, the morality of stem cell extraction may no longer be an issue

for researchers.

Table 1. Possible therapeutic uses of tissue derived from stem cells	Lisker (608)
Cell type	**Diseases**
Neural	Stroke, Parkinson's, Alzheimer's, Spinal cord injury
Heart muscle	Coronary, Congestive heart failure
Insulin-producing	Diabetes
Blood	Leukemia, Genetic blood diseases
Liver	Hepatitis, Cirrhosis
Skin	Burns
Retinal	Macular degeneration
Skeletal muscle	Muscular dystrophy

Dr. Haruko Obokata and her colleagues conducted a recent study in which

they developed a new method of obtaining pluripotent stem cells. The

technique involved submerging white blood cells from a baby mouse in a mild

acid for approximately thirty minutes. The pH of the acid was about 5.7, or a

little more acidic than a gallon of milk. After a few days, the cells stopped behaving like blood cells, and started behaving like stem cells. Subsequently, the scientists injected the cells into a mouse embryo and the cells acted just like all the other stem cells. The research group has named these cells "Stimulus-Triggered Acquisition of Pluripotentcey", or STAP. The team said that they aren't exactly sure how it works, as it is quite a messy process. Dr. Obokata and her team are still testing to see if the method works with adult mice as well as in humans. If these scientists had continuous support they needed for their research, they could complete experiments at a faster pace, as well as establish cloning methods that do not involve embryonic material.

Despite the bad connotations that therapeutic cloning and reproductive cloning have, both contribute to the advancements of public health and biotechnology. As mentioned before, therapeutic cloning has potential to grow organs or tissue that are genetically identical to a patient; thus producing a decrease in rejection and organ failure. Similarly, other procedures such as breast implants or cosmetic surgery can cause immune diseases; but if the doctor used genetically identical material to rebuild a patient's tissue, the immune deficiencies caused by surgeries would soon be over. On the other hand, reproductive cloning methods would have the ability to genetically produce a child or living organism from two parents. For example, if a couple was unable to produce children, they could pay for a procedure that would extract genetic material from each parent, then use that material to create a child (Shih 30). However, this procedure would cause an economical imbalance as the less fortunate families would not be able to produce genetically identical children; leaving the procedure only for the richer families. This technique would have to be cost-effective enough to where anyone could pay for the procedure. With funding, therapeutic and reproductive cloning could become cost effective, allowing the benefits of cloning to be introduced to public health techniques.

In conclusion, therapeutic cloning and reproductive cloning are both extremely controversial due to society's misconception of the definition of cloning. As mentioned, there have been legal complications, religious complications and research complications throughout the years; yet, scientists continue to explore the world of replication. Moreover, the misunderstanding of the definition of cloning has caused a major decrease in support as well as funding. This technique should not be removed due to the influence of a few organizations. Researchers must continue to press on and spread the word of therapeutic cloning and reproductive cloning as well as all of their potential. Therefore, with the support and funds of society, scientists and researchers can work to establish cloning methods that do not involve embryonic material, thus satisfying the legislative and religious dispute.

Works Cited

Bailey, Regina. "Bacterial Reproduction." *About.com Biology*. N.p., n.d. Web. 20 Feb. 2014.

Nikas, Nikolas, Dorinda Bordlee, and James Hanson. "Human Cloning Laws: 50 State Survey." *Bioethics Defense Fund*. Web. 2 February 2014.

Hansen, Brian. "Cloning Debate." *CQ Researcher*. 22 Oct. 2004. Web. 11 Feb. 2014.

Roleff, Tamara. <u>Cloning: Opposing Viewpoints</u>. Farmington Hills, MI: Green Haven Press, 2006.

Never let me go. Dir. Mark Romanek. Perf. Andrew Garfield, Keira Knightly, Carey Mulligan. Twentieth Century Fox Home Entertainment, 2010. DVD.

Orphan Black. Manson, Graeme, and John Fawcett. BBC America. 2013. Television.

Veselka, Camille. "A Detrimental Influence: The Effect Religion Has on Laws." *The Huffington Post*. TheHuffingtonPost.com, 5 Dec. 2011. Web. 9 Mar. 2014.

Weasel, Lisa and Eric Jensen. "Language and values in the human cloning debate: a web-based survey of scientists and Christian fundamentalist pastors." *New Genetics and Society, Vol. 24, No. 1*. April 2005. Web. 4 February 2014.

"CREDONG.org." Largest Religions: Christianity -. N.p., n.d. Web. 11 Mar. 2014.

Doucleff, Michaeleen. "A Little Acid Turns Mouse Blood Into Brain, Heart And Stem Cells." *WNCW*. 30 Jan. 2014. Web. 2 Feb. 2014.

Gomes, Jaquetta. "The development and use of the Eight Precepts for lay practitioners, Upaˉsakas and Upaˉsikaˉs in Theravaˉda Buddhism in the West." *Contemporary Buddhism, Vol. 5, No. 1*. 2004. Web. 4 February 2014.

Larijani, B, and F Zahedi. "Islamic Perspective On Human Cloning And Stem Cell Research." Transplantation Proceedings 36.10 (2004): 3188-3189. Print.

Shih, Ching-Pou. "Moral and Legal Issues Concerning Contemporary Human Cloning Technology." *Golden Gate University School of Law*. N.p. Web. 2 February 2014.

Lisker, Rubén. "Ethical and Legal Issues in Therapeutic Cloning and the Study of Stem Cells." *Elsevier*. 19 June. 2003. Web. 2 February 2014. Table.

Smith, Simon. "The Benefits of Human Cloning." The Benefits of Human Cloning. N.p., 26 Feb. 1998. Web. 25 Mar. 2014.

Starfish from page 4; Image citation:

N/A. *Starfish*. N.d. What can Michael Jordan teach us about the body?, bodyLITERATE
Research Initiative. *Body Literate*. Web. 18 Aug. 2014.

ABOUT THE AUTHOR

I am a young student from Wilmington, North Carolina aspiring to be a biologist. I currently attend Isaac Bear Early College High School, and I am taking all of my courses full time at UNCW. I am very passionate about Biology and I hope to pursue a career in the field of Microbiology. I plan to transfer to North Carolina State University in the Fall of 2015. There, I will obtain a bachelor's degree in biology, as well as a master's degree in microbiology. I would like to conduct a research study for a few years once I receive my master's degree. Overall, I am a diligent worker who loves the world of science and I hope to continue to learn as I grow.

www.ingramcontent.com/pod-product-compliance
Lightning Source LLC
Chambersburg PA
CBHW050409180526
45159CB00005B/2207